YOUR KNOWLEDGE HAS VALUE

- We will publish your bachelor's and
  master's thesis, essays and papers

- Your own eBook and book -
  sold worldwide in all relevant shops

- Earn money with each sale

Upload your text at www.GRIN.com
and publish for free

Bibliographic information published by the German National Library:

The German National Library lists this publication in the National Bibliography; detailed bibliographic data are available on the Internet at http://dnb.dnb.de .

This book is copyright material and must not be copied, reproduced, transferred, distributed, leased, licensed or publicly performed or used in any way except as specifically permitted in writing by the publishers, as allowed under the terms and conditions under which it was purchased or as strictly permitted by applicable copyright law. Any unauthorized distribution or use of this text may be a direct infringement of the author s and publisher s rights and those responsible may be liable in law accordingly.

Imprint:

Copyright © 2017 GRIN Verlag, Open Publishing GmbH
Print and binding: Books on Demand GmbH, Norderstedt Germany
ISBN: 9783668471573

This book at GRIN:

http://www.grin.com/en/e-book/367068/reliability-centered-maintenance-reliability-engineering-and-asset-risk

Salisu Alhassan

Reliability Centered Maintenance. Reliability Engineering and Asset Risk Management

GRIN Publishing

GRIN - Your knowledge has value

Since its foundation in 1998, GRIN has specialized in publishing academic texts by students, college teachers and other academics as e-book and printed book. The website www.grin.com is an ideal platform for presenting term papers, final papers, scientific essays, dissertations and specialist books.

Visit us on the internet:

http://www.grin.com/

http://www.facebook.com/grincom

http://www.twitter.com/grin_com

ASSIGNMENT ON

RELIABILITY ENGINEERING AND ASSET RISK MANAGEMENT

By:

SALISU SALIHU ALHASSAN

21/01/2017

<u>**ANSWER TO QUESTION 1.**</u>

Serial connection Reliability block diagram (RBD) of a radar system.

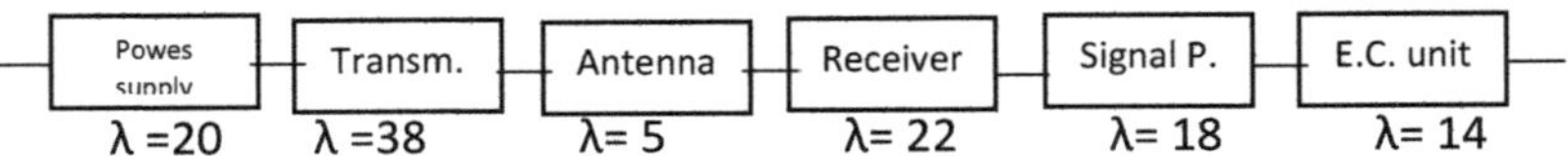

$\lambda = 20$ $\lambda = 38$ $\lambda = 5$ $\lambda = 22$ $\lambda = 18$ $\lambda = 14$

a) To calculate for MTBF,

    Using: MTBF = MTTF + MTTR

    And MTTF = $1/\lambda$

    $\lambda_T = 20+38+5+22+18+14 = 117 * 10^{-6}$ hrs

    $\therefore$ MTTF = $1 / 117 * 10^{-6}$

    MTTF = <u>8,547hrs</u>

For a real system, since repair time is minimal, then

    MTBF = MTTF

    MTBF = <u>8,547hrs</u>

Below is the Analysis using a Simulation Model:

	20	38	5	22	18	14	λ_T =0.000117
	R	X					
	0.523967	5524.164					
	0.273427	11083.09					
	0.642567	3780.212					
	0.708734	2942.525					
	0.496781	5979.541					
	0.742386	2546.037					
	0.996691	28.32468					
	0.507045	5804.744					
	0.538867	5284.498					
	0.545219	5184.333					
	0.130445	17408.58					
	0.815911	1738.892					
	0.876644	1125.248					

0.392494	7993.461					
0.156523	15850.87					
0.755742	2393.633					
0.632611	3913.677					
0.263455	11400.61					
0.001438	55937.91					
0.563558	4901.576					
	8541.097hrs	←	**MTBF**			

b) The reliability target of achieving 10,000hrs MTTF for the system

$$MTTF = 1/\lambda$$

$$\therefore \lambda = 1/MTTF \quad \rightarrow \quad \lambda = 1/10,000hrs$$

$$\lambda = 100*10^{-6}hrs$$

Now calculating to get $\lambda_T \equiv 100$ can be achieved by adding redundancy to the transmitter

$$\therefore \lambda_T = 20 + (38/2) + 5 + 22 + 18 + 14$$

$$\lambda_T = \underline{98*10^{-6} \ hrs}$$

Reason for adding the redundancy to the transmitter is because it has the higher failure rate.

ANSWER TO QUESTION 2.

CONSIDERING THE FIRST SENSOR

The probability of a fault occurring = 0.02

Probability of detecting the fault = 0.95

The probability of false alarm = 0.10

Let A = Report Fault

Let B = Actual fault

Let B/A should be that a reported fault has actually occurred.

Then the probability that a fault has actually occurred and is reported by the first sensor is = P (B/A)

Using Bayes' Theorem:

P (B/A) = P (A/B) * P(B)/P(A/B)*P(B) + P(A/B')* P(B)

P (B/A) = 0.02 * 0.95/ (0.02*0.95) + (0.1*0.98)

P (B/A) = 0.019/0.117

P (B/A) = 0.1624

CONSIDERING THE SECOND SENSOR

The Probability of a fault occurring = 0.02

The Probability of detecting a fault is = 0.90

The Probability of false alarm = 0.05

Let A = Report Fault

Let B = Actual fault

Let B/A be that a reported fault has actually occurred

 Then, the probability that a fault has actually occurred and is reported by the second sensor = P (B/A)

Using Bayes' Theorem:

P (B/A) = P (A/B) * P(B)/P(A/B)*P(B) + P(A/B')* P(B)

P (B/A) = 0.02 * 0.90/ (0.02*0.90) + (0.05*0.98)

P (B/A) = 0.018/0.067

P (B/A) = 0.2687

ANSWER TO QUESTION 3

	USING X= a+(b-a)*R			a= 0	b= 8	R= random numbers	
	R	x1		R	x2		Y
	0.182094	1.456756		0.109767	0.878137		1.456756
	0.670761	5.36609		0.676967	5.415732		5.415732
	0.509713	4.077703		0.380395	3.043158		4.077703
	0.437573	3.500588		0.008248	0.065987		3.500588
	0.083342	0.666739		0.486694	3.893553		3.893553
	0.533865	4.270922		0.16193	1.295439		4.270922
	0.162971	1.30377		0.698778	5.59022		5.59022
	0.375721	3.005772		0.379728	3.037822		3.037822
	0.802467	6.419737		0.461098	3.688787		6.419737
	0.944423	7.555384		0.664577	5.31662		7.555384
	0.607247	4.85798		0.061967	0.495733		4.85798
	0.022538	0.180307		0.371946	2.975564		2.975564
	0.19242	1.539362		0.975067	7.800539		7.800539
	0.890969	7.127754		0.123752	0.990013		7.127754
	0.198565	1.588518		0.43708	3.496642		3.496642
	0.851645	6.813158		0.103899	0.831195		6.813158
	0.124701	0.997608		0.720208	5.761662		5.761662
	0.184928	1.479424		0.792042	6.336336		6.336336
	0.986391	7.891127		0.912862	7.302894		7.891127
	0.523503	4.188022		0.204415	1.635324		4.188022
	0.401178	3.209423		0.687593	5.500743		5.500743
	0.507878	4.063025		0.880044	7.040351		7.040351
	0.081924	0.655396		0.171243	1.369945		1.369945
	0.076052	0.608414		0.868114	6.94491		6.94491
	0.597759	4.782069		0.139088	1.1127		4.782069
	0.546305	4.370436		0.016931	0.135445		4.370436
	0.270245	2.16196		0.079821	0.638567		2.16196
	0.569479	4.555828		0.593483	4.747868		4.747868
	0.78722	6.297757		0.617468	4.939747		6.297757
	0.339489	2.715911		0.906883	7.255066		7.255066
	0.754968	6.039742		0.587834	4.702673		6.039742
	0.139976	1.119809		0.763835	6.110679		6.110679
	0.380099	3.040792		0.308077	2.464617		3.040792
	0.298847	2.390778		0.288665	2.309318		2.390778
	0.700673	5.605387		0.099512	0.796093		5.605387
	0.240588	1.924707		0.896516	7.172128		7.172128
	0.366243	2.929945		0.488052	3.904415		3.904415
	0.149399	1.195194		0.404905	3.239238		3.239238
	0.782767	6.262134		0.682677	5.461417		6.262134
	0.073301	0.586408		0.852187	6.817499		6.817499

	0.333701	2.669604		0.805013	6.440101		6.440101
	0.449659	3.597269		0.259899	2.079192		3.597269
	0.325619	2.604955		0.660566	5.284528		5.284528
	0.694944	5.559552		0.014268	0.114146		5.559552
	0.573547	4.588376		0.175814	1.406512		4.588376
	0.40212	3.216964		0.59593	4.767444		4.767444
	0.78159	6.252718		0.394317	3.154538		6.252718
	0.885633	7.085065		0.701273	5.610183		7.085065
	0.312114	2.496914		0.395746	3.165968		3.165968
	0.210099	1.680789		0.219409	1.755273		1.755273
	0.705737	5.645899		0.422121	3.376965		5.645899
	0.292177	2.337412		0.621246	4.969968		4.969968
	0.454385	3.635077		0.691233	5.52986		5.52986
	0.954288	7.634306		0.589405	4.715237		7.634306
	0.48693	3.895441		0.396153	3.169224		3.895441
	0.328255	2.626038		0.79024	6.32192		6.32192
	0.574693	4.597546		0.978772	7.830176		7.830176
	0.610435	4.883482		0.964819	7.718551		7.718551
	0.746572	5.972576		0.990028	7.920227		7.920227
	0.405002	3.240018		0.410402	3.283216		3.283216
	0.962662	7.701293		0.5281	4.224802		7.701293
	0.521805	4.174442		0.867096	6.936772		6.936772
	0.581726	4.653811		0.252047	2.016377		4.653811
	0.506568	4.052542		0.84705	6.776398		6.776398
	0.729411	5.835289		0.579304	4.634435		5.835289
	0.68647	5.491762		0.673369	5.386952		5.491762
	0.41681	3.33448		0.382892	3.063139		3.33448
	0.425596	3.404769		0.677164	5.417311		5.417311
	0.080415	0.643321		0.705753	5.646021		5.646021
	0.716209	5.729674		0.150687	1.205497		5.729674
	0.739817	5.918534		0.347573	2.780584		5.918534
	0.858102	6.864813		0.066369	0.530952		6.864813
	0.039762	0.318094		0.742996	5.943971		5.943971
	0.495282	3.962257		0.497908	3.983261		3.983261
	0.229883	1.839064		0.039227	0.313819		1.839064
	0.729358	5.834863		0.315329	2.522635		5.834863
	0.807843	6.462741		0.550749	4.405989		6.462741
	0.037558	0.300463		0.960876	7.68701		7.68701
	0.304925	2.439399		0.620173	4.961384		4.961384
	0.96382	7.710562		0.685491	5.48393		7.710562
	0.373114	2.98491		0.421686	3.373488		3.373488
	0.604998	4.83998		0.308065	2.464518		4.83998
	0.234955	1.879643		0.134139	1.47311		1.879643
	0.658454	5.267635		0.999917	7.999334		7.999334

	0.217304	1.738434		0.793872	6.350977		6.350977
	0.584783	4.678262		0.004971	0.039767		4.678262
	0.729673	5.83738		0.566825	4.534604		5.83738
	0.72456	5.796479		0.61917	4.953362		5.796479
	0.172607	1.380858		0.340347	2.72278		2.72278
	0.041972	0.335774		0.297999	2.383989		2.383989
	0.226379	1.811035		0.750993	6.007948		6.007948
	0.473634	3.78907		0.679844	5.438748		5.438748
	0.606707	4.853654		0.207048	1.656387		4.853654
	0.193952	1.551617		0.842437	6.739493		6.739493
	0.576045	4.608357		0.255532	2.044252		4.608357
	0.352351	2.818805		0.293099	2.34479		2.818805
	0.274661	2.197286		0.750718	6.005743		6.005743
	0.856439	6.851512		0.18566	1.485284		6.851512
	0.255778	2.046224		0.515206	4.121646		4.121646
	0.085785	0.68628		0.292462	2.339696		2.339696
						MTTF→	**5.231439**

RELIABILITY-CENTERED MAINTENANCE AND ITS APPLICATIONS TO OFFSHORE WIND TURBINES

1. INTRODUCTION

For so many years, maintenance have been experiencing trends of transitions from one stage to another. These stages of transformational procedures are due to the fact that all assets that are put into operation requires essential maintenance so as to improve in the availability of the system and also to achieve efficient and effective functionality of the system. Maintenance must be done to ensure a system is available and functioning well thus utilising the system optimally and minimising cost (Moubray,1997). This paper is going to focus on a maintenance method that provides reliability to a functional system and in a cost-effective manner known as Reliability centered maintenance technique. According to Moubray (1997) RCM is "a procedure that is used to ascertain what should be done to make sure that all component of a system continues to perform its intended function and in the way its user wants it to operate in its current operating circumstance"

RCM is simply referred to as the idea of considering the lasting reliability of system. It includes coming up with ideas and ways of maintaining the system, and to make sure it is reliable throughout its expected life period. It encompasses adding weight to the choice of systems which is recognised to be dependable and sustainable and for which logistic sustenance is most gladly delivered. In practice this often means choosing systems that are readily accessible off-the-shelf and which are since commonly used. It also includes examining for reliability and acceptable installation at the time of acquiring the asset. Key issues are system harmony, reliability and maintainability, assessment, and acceptance testing (Nicholas A.J,2010).

2. HISTORY AND EVOLUTION OF RCM

Maintenance task was prepared by the civil aircraft sector of the United State in 1974, a duty assigned to the sector by the United State defence department. The sector had been terrified by cost of maintenance over a long period of time, and that is what led to the investigation that brought about the bank of idea of Reliability centered maintenance for the aircraft producers as a tool to carry out maintenance of their aircraft. This idea pave way for the 747-maintenance steering group document. It was the first applicable program for the Boeing aircraft. Also, a revised form of the document was applied to Lockleed L1011 and Douglas DC- 10. In summary, the whole idea of RCM was documented and published in 1978 by Stan Nowlan and Howard Heap, proven to be a document upon which the knowledge of reliability centered maintenance is relied upon. Finally, Nowlan and Heap (1978) found out that a lot of failures are inevitable irrespective of the maintenance that is done, and it was found pertinent noting that for several components, the probability of a component to fail is not increasing with age.

3.0   PROCESSSES OF RCM

RCM today is playing a vital role in the Offshore wind turbine sector by reducing the cumbersome amount of cost of maintenance terrifying the sector. RCM works hand in hand with the traditional maintenance processes to achieve reliability in a cost-effective manner (Nowlan and Heap,1978).

3.1   TECHNIQUES INTEGRATED WITH RCM

There are four processes integrated with RCM technique:

(i)   Reactive method

(ii)   Preventive method

(iii)   Predictive testing and inspection method

(iv)   Proactive method respectively.

The target of RCM is to recognise the most cost-effective and elated maintenance method to minimise risk and to mitigate failure in system. This permits systems operability to be sustained in the most cost-effective means (Nowlan and Heap,1978).

3.2   STRUCTURE OF RCM

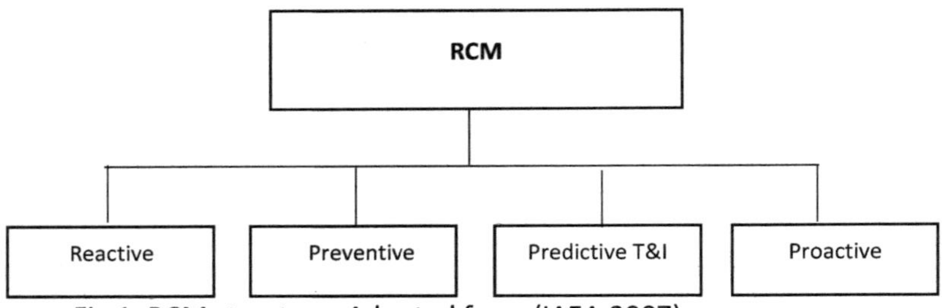

Fig 1: RCM structure. Adopted from (IAEA,2007).

3.3   OBJECTIVES OF RCM

One important thing to note about RCM is that it integrates completely all the above listed investigation results, as no one is more significant than the other (IAEA,2007).

(i)   Make no Maintenance: This means Reactive maintenance, restoration, repair when it fails, or keep using it until it fails. This kind of maintenance assumes that failure is similarly possible to arise at any component, and that the failure is non-damaging to the process. With this type of maintenance, high cost and risk are the consequences.

(ii)   Performing Preventive Maintenance: this consists of often scheduling inspections, adjusting, lubricating, and replacing of failed components. Preventive maintenance is also termed as interval based maintenance or time driven maintenance. Done without taking into consideration the component state. Preventive maintenance plans check-up and maintenance at predefined intervals to decrease system failures. Nowlan and Heap revealed that Preventive Maintenance can lead to a substantial rise in inspections and cost without any improvement in reliability.

(iii)   Performing Condition Based Maintenance: this method comprises of Predictive Maintenance and real-time monitoring. Condition based maintenance replaces subjectively scheduled maintenance tasks by maintenance which is planned only when necessarily it is required by condition of the system. Continual study of the asset condition permits for a planned and scheduled maintenance actions for before any major failure occurs.

(iv)   Redesigning the system: When any of the three-listed maintenance process is not favourable, redesigning the system helps in makes the system equivalent to a new one, by the addition of a redundant system thereby doing away with the possibility of any risk that might occur and in a cost-effective manner.

3.4   PRINCIPLES OF RCM

1. RCM is more concern with the whole asset as a system than the parts of the system.

2. RCM is motivated by safety and economics because operators will always want to be sure that his system is safe and will not cause any form of damage, loss of life and loss of property to ensure cost effectiveness.

3. RCM ensures apart from system being efficient, the user should also be able to control the performance of the system.

4. RCM considered the issue of discomfort. There should be comfort, and Reliability centered maintenance has recognised the need for ageing investigation processes.

5. RCM is flexible with environmental integrity by blending with them in an appropriate manner.

6. RCM also considered the appearance of system as well as projecting the image of the asset. Example; painting to avoid corrosion

7. RCM also reduces the way a failure might occur using protective device, which plays some important roles like:

    - Sensors and Alarms that calls the attention of the user when there is a problem
    - To put off the system when a failure has occurred
    - To have safety critical elements in place in case of emergencies
    - Redundancy elements to serve as a backup
    - To stop hazardous conditions from occurring in the first instance

8 RCM identifies the way an equipment or the system failure can occur

9 RCM uses information from the output of the logic tree to increase in the reliability of the system.

10 RCM improves reliability and cost benefits.

11 RCM uses a logic tree to assess maintenance, which can virtually be applied to all type of systems.

3.5   RCM METHODOLOGY

According to Moubray (1997) RCM is "attributed with seven (7) key questions"

1. What are the functions of the system?

2. What operational failures are probable to occur?

3. What are the reasons for the failures?

4. What is the repercussion when a failure occurs?

5. How does the failure occur?

6. What can be done to predict or terminate failures?

7. What should be done to reduce the possibility of failure or its consequences?

3.6   OUTCOME OF RCM ANALYSIS

If RCM is applied properly according to the guide of logic tree, it gives an RCM analysis result in three concrete outcomes (Moubray,1997).:

1. Maintenance task should be carried out by the maintenance team

2. Reviewed operation processes for the system operators

3. Comprehensive list of the part of the system that requires attention must be made and take necessary actions to ensure the asset performs its function within its expected context.

4. Failure mode effects and criticality analysis (FMECA): The working life of the system is assessed by FMECA which interprets and documents functional failures, failures, failure modes and the effects of the failures at different stages and finally rank these condition as a tool for assessing the system.

Fig 2: RCM Logic tree, Adopted from (IAEA,2007).

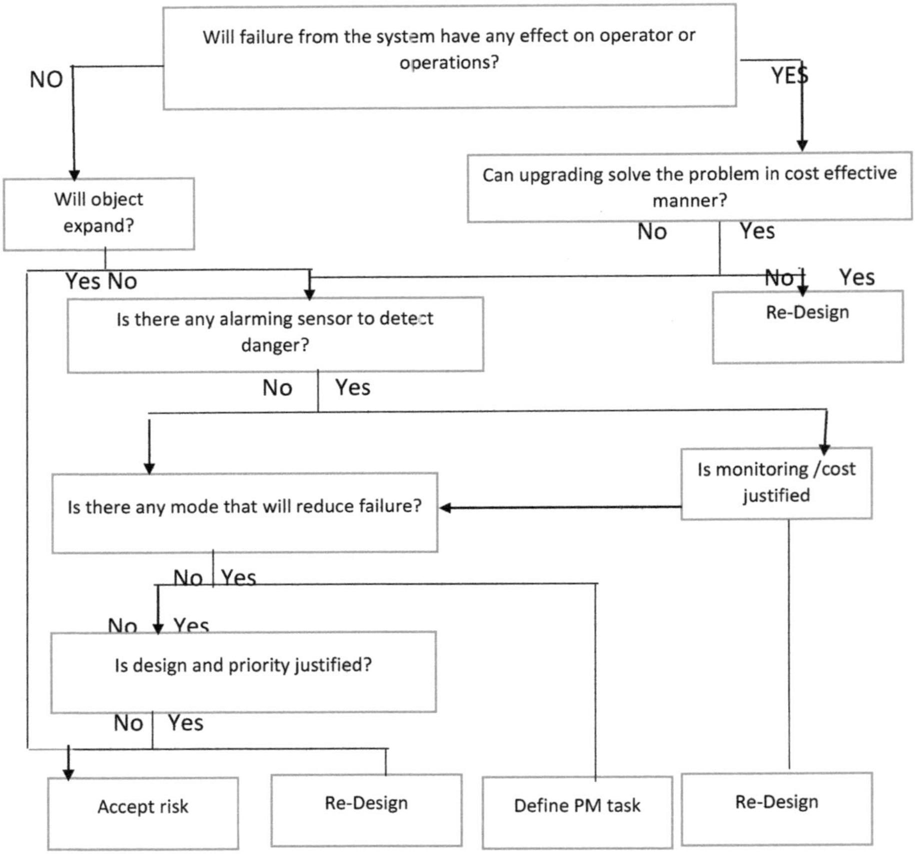

Will failure from the system have any effect on operator or operations?

NO

YES

Will object expand?

Can upgrading solve the problem in cost effective manner?

No   Yes

Yes No

Is there any alarming sensor to detect danger?

Not   Yes

Re-Design

No | Yes

Is there any mode that will reduce failure?

Is monitoring /cost justified

No | Yes

No | Yes

Is design and priority justified?

No | Yes

Accept risk

Re-Design

Define PM task

Re-Design

4.0   APPLICATION OF RCM TO WIND TURBINES

RCM technique if properly applied, tells the operator on the way the offshore wind turbine system in service is bound to experience a failure and the subsequent severity that will follow. To achieve a standard maintenance approach, RCM is integrated with other traditional methods to achieve reliability of the wind turbine system. According to Manwell J. et al (2006) The prime purpose of wind turbine mechanism is to converts wind power energy into electricity. Wind turbine is characterised by various Failure modes, but application of RCM helps in mitigating these effects by following some decisive and active phases (Luengo and Kolios, 2015).:

i.   Classifying the wind turbine functionality into critical stages

ii.   It then identifies and ranks the failure modes of the wind turbine system.

iii. Finally, following the RCM basic principles in implementing the approach.

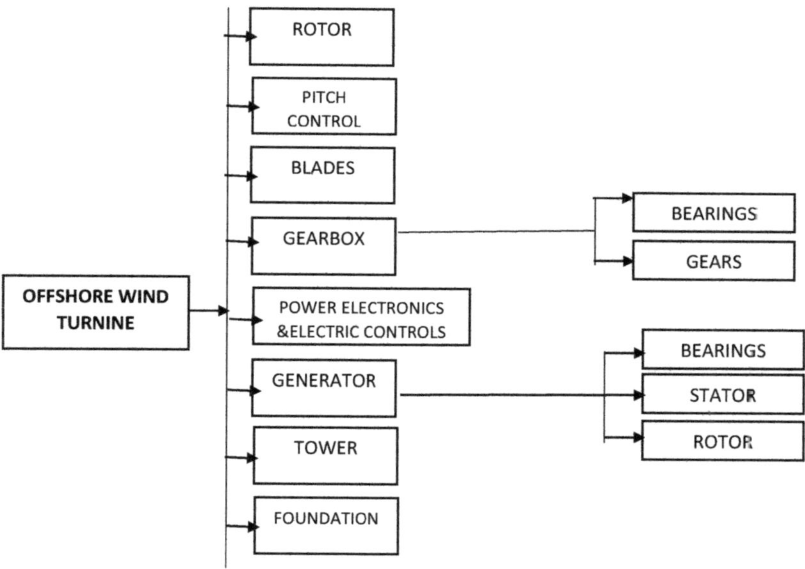

Fig 3. Wind turbine and its components. Adopted from (Luengo and Kolios, 2015)

The take-home idea is that a decrease in risk of failure is experienced by the restoration and whole maintenance processes. However, after a transitory increase in operation costs, the offshore wind turbine farm operator will realise an appreciable reduction in total operation expenditure (Moubray,1997).

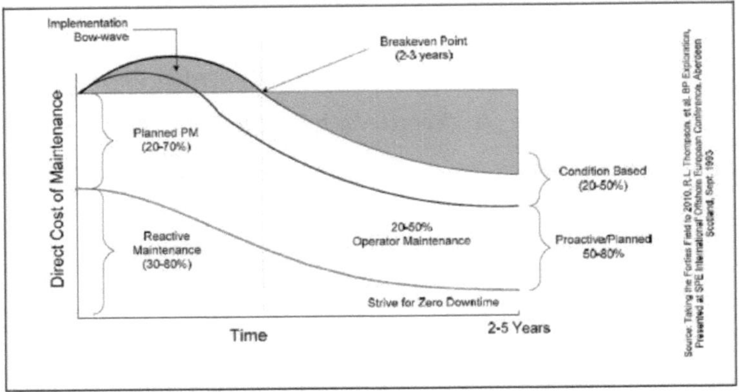

Figureadoptedfrom:http://www.upwindsolutions.com/blog/bid/340036/reliability-centered-maintenance-in-wind-farm-operations

4.1 BENEFITS OF RCM

a) Improves Reliability: Reliability procedure must not only be to decrease the cost of maintenance but also to increase in the operability of the system by ensuring it functions optimally. Hence, improved reliability and efficiency of the wind turbine will in return contribute to improved economic and safety of the system.

b) Contribution to Long Term Operation: RCM is an organized method of identifying and executing maintenance tasks intended for supporting the performance of the wind turbine.

c) Better maintenance cost benefit

d) Longer useful life of exclusive components

e) Emphasis on record keeping

f) Educating people involved in the maintenance processes so they can have idea on how their asset functions and how to proffer solutions when carrying out maintenance.

g) RCM also encourages team work. (IAEA,2007).

4.2   LIMITATIONS OF RCM

According to Moubray (1997) RCM has some disadvantages in the cause of its application:

a) RCM is time consuming
b) Training is required for the people involved in the process
c) RCM Focus too much on failure data
d) RCM is not computer friendly
e) RCM is most at times monopolised by maintenance department

4.3   CONCLUSION

RCM has been proved to be a unique maintenance technique that keep eyes on what the intended function of a system was, although a time-consuming method that considers critical elements but, has the ability of detecting failures even before it occurs and tries to proffer suitable preventive maintenance action thereby increasing in the reliability of the asset in a cost-effective way.

REFERENCES

1. Moubray, J., 1997. *''RCM- Reliability centered maintenance''*, second edition, Industrial press Inc. 989 Avenue New York, United State of America.
2. Nicholas, J., 2010. *"Physical Asset Management"*, Springer London Dordrecht Heidelberg.
3. Manwell J.F, McGowan J.G and Rogers A.L 2006. *'Wind Energy Explained, Theory, Design and Application.,* The Atrium, Southern Gate, Chichester, West Sussex, England.
4. Nowlan, F. and Heap, F., 1978. *"Reliability Centered Maintenance''*, Published by United Airlines Washington D.C., U.S.A
5. International Atomic Agency., *''Application of Reliability Centered Maintenance to Optimise Operation and Maintenance in Nuclear power plant''*, IAEA Wagramer Strasse, Vienna, Austria. (2007).
6. Luengo, M.M. and Kolios, A. (2015) *'Failure Mode Identification and End of Life Scenarios of Offshore Wind Turbines: A Review'*, , pp. 8339–8354.

7. Htttp://www.upwindsolutions.com

YOUR KNOWLEDGE HAS VALUE

- We will publish your bachelor's and master's thesis, essays and papers

- Your own eBook and book - sold worldwide in all relevant shops

- Earn money with each sale

Upload your text at www.GRIN.com and publish for free